creatures
of the sea

Squid

Other titles in the series:

Creatures of the sea

Squid

Kris Hirschmann

KIDHAVEN
PRESS™

San Diego • Detroit • New York • San Francisco • Cleveland
New Haven, Conn. • Waterville, Maine • London • Munich

For more information, contact
KidHaven Press
27500 Drake Rd.
Farmington Hills, MI 48331-3535
Or you can visit our Internet site at http://www.gale.com

LIBRARY OF CONGRESS CATALOGING-IN-PUBLICATION DATA

Hirschmann, Kris, 1967-
   Squid / by Kristine Hirschmann.
     p. cm. — (Creatures of the sea)
Summary: Describes the physical characteristics, behavior, predators, and life cycle of the squid.
Includes bibliographical references (p.   ).
  ISBN 0-7377-1556-1 (hardcover: alk. paper)
  1. Squids—Juvenile literature. [1. Squids.] I. Title.
QL430.2.H57 2004
594'.58—dc22
                                                              2003014091

Table of Contents

Introduction

Sea Monsters?

In the twelfth century, Norwegian sailors believed that a squidlike monster called the Kraken roamed the world's oceans. According to the sailors' stories, the Kraken had many wriggling arms. It was so big that it could reach the top of a sailing ship's mast. Once the Kraken had grabbed a ship, it would pull it over and eat the crew.

Centuries later, author Jules Verne published a book called *20,000 Leagues Under the Sea*. In this book a group of enormous squid attacks a submarine. To save themselves, the crew uses axes to hack off the arms of these squid. One man dies when he is pulled into the sea by an especially large beast.

Stories like these reflect the human fascination with squid. These creatures are common throughout the world's oceans. They can be found everywhere, from the cold seas near the poles to the warm waters of the equator. Because squid are so common, it is no surprise that people have been bumping into them since the beginning of recorded history.

Scary squid stories probably come from human contact with just a few "monstrous" species. The Humboldt squid, for example, may weigh three hundred pounds and can grow to eight feet in

Although stories about giant squid are popular, most squid are small like this one.

length. The giant squid is another huge species. Scientists believe that these animals can reach lengths of one hundred feet.

The Humboldt squid and the giant squid are exceptions. There are about three hundred squid species, and most are less than two feet long. Far from being sea monsters, they are interesting creatures that play an important role in the ocean food chain. Today, squid—even the big and fierce ones—are studied rather than feared. Scientists and sailors alike have come to appreciate these amazing creatures of the sea.

Sea Arrows

Squid belong to a group of animals called cephalopods. The word *cephalopod* comes from two Greek words meaning "head" and "foot." It describes animals whose feet grow directly from their heads. This group includes octopuses, cuttlefish, and nautiluses as well as squid.

Most squid have long, pointed bodies. For this reason, they are sometimes called "sea arrows." The nickname is a good one. Squid zip through the world's oceans like arrows shot from a hunter's bow.

The Mantle and Head

The arrowlike part of the squid's body is called the **mantle**. The mantle contains the squid's internal organs. Its outer layer is made of smooth skin. Just beneath the skin is a thick layer of muscle, which is attached to a stiff structure called the **pen**. The pen is made of **chitin**, a material that is similar to human fingernails. It is the closest thing the squid has to bones.

Two fins are attached to the mantle's tip. The fins keep the squid stable in the water as it swims. They can also wave to hold the squid motionless or wiggle gently to help the squid make careful movements.

Squid are sometimes called "sea arrows" because of their long, pointed bodies. This Caribbean reef squid is one of about three hundred varieties.

Just below the mantle is the squid's head. The head is much smaller than the mantle, and it is joined to the mantle by a short neck. The head contains the squid's two large eyes. It also contains the mouth, which is hidden between the squid's arms and **tentacles**.

Arms and Tentacles

Attached to the base of the head are eight rubbery arms. The arms contain no bones and are very flexible. Depending on the species, the arms can be long and thin, short and stubby, or any size and shape in between.

In all species, the inner surface of each arm is lined with round disks called **suckers**. The squid uses its suckers to grab and hold things. It does this by placing its suckers against an object, then using muscles to pull up on the suckers' centers. This creates suction and makes the suckers cling to anything they touch. Many squids' suckers are also lined with hooks or "teeth" made of chitin. These sharp hooks dig into soft objects, giving the squid an even better grip.

In addition to their eight arms, squid also have two tentacles. The tentacles are longer than the arms and consist of a stalk and a club. The stalk is long and thin, and it has no suckers. The club is a flat, wide part at the end of the stalk, and it is covered on one side with small suckers.

This close-up of the club at the end of a squid's tentacle shows the suckers. The teeth on the sucker rings help the squid hold onto struggling prey.

Getting Around

Squid do not use their arms and tentacles to swim. Instead, they get around in their ocean homes by waterpower. A squid sucks water into its **mantle cavity**, which is just beneath the outer skin of the mantle. It uses fleshy knobs to "lock" its mantle and head together, then it squeezes the mantle. This action forces water out through a tube called the **siphon** or **funnel**. The blasting water pushes the squid through the sea. A squid blasts water from its funnel over and over to get wherever it needs to go.

Squid use their funnels to control their direction. How? A squid points its funnel anywhere it likes to change the direction of its water blast. This in turn changes the squid's direction. By pointing its funnel in different ways, a squid can swim forward, backward, sideways, up, or down without turning its body.

A squid can also use its funnel to control its speed. By opening the funnel wide, a squid makes its water blast less powerful and therefore slows itself down. By squeezing the funnel into a narrow tube, the squid produces a stronger water blast and swims more quickly.

Built for Speed

Squid have another feature that helps them to swim quickly. These creatures have the longest, thickest

A school of squid swims through the ocean. Some squid can move as fast as twenty miles per hour for brief periods.

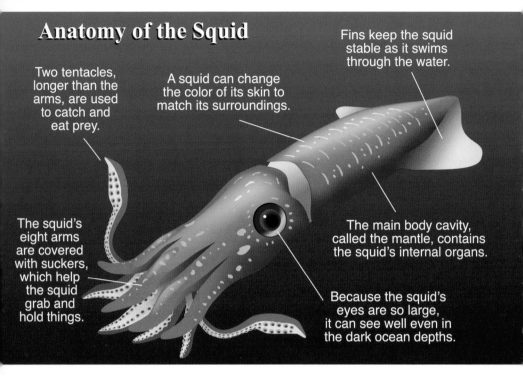

Anatomy of the Squid

Fins keep the squid stable as it swims through the water.

Two tentacles, longer than the arms, are used to catch and eat prey.

A squid can change the color of its skin to match its surroundings.

The squid's eight arms are covered with suckers, which help the squid grab and hold things.

The main body cavity, called the mantle, contains the squid's internal organs.

Because the squid's eyes are so large, it can see well even in the dark ocean depths.

nerves of any animals in the world. A squid's biggest nerves, sometimes called **giant axons**, run all the way from one end of the body to the other. They may be a hundred times thicker than human nerves.

In all animals, nerves are the body's "wires." They carry electrical messages from the brain to every part of the body. Electrical messages travel much faster through long, wide nerves than short, thin ones. Because the squid has such big nerves, its brain can instantly send messages to every part of the body. This means a squid can respond to any situation very quickly.

Big nerves also allow a squid to squeeze every part of its mantle at the same instant. This simultaneous push creates a lot of pressure and forces water out through the funnel at a very high speed. As a result, the squid moves very quickly. Some squid can swim as fast as twenty miles per hour for brief stretches. This speed qualifies squid as the world's fastest **invertebrates** (animals without backbones).

Changing Color

Squid have another use for their amazing nervous systems. They use them to change the color of their skin, thereby **camouflaging** themselves. To *camouflage* means to blend into the background and become very hard to see. Squid do this by changing their color to look like what is around them.

Squid use special cells called **chromatophores** to change their color. Chromatophores come in many different colors, including red, yellow, brown, and black. Each chromatophore is controlled by small muscles. When the muscles are relaxed, the chromatophore is tiny. But when the muscles contract, the chromatophore is pulled open. Its color spreads over a larger area. A squid controls its skin color by pulling open certain colors and patterns of cells.

Squid have very good control over their chromatophores. Most species can make between thirty and fifty different patterns, and they can completely change their color in about one-third of a second. Some squid can even "ripple" their colors to blend with shifting sunbeams or blink skin patterns on and off. Blinking their skin patterns is one way squid communicate with other squid.

Squid Senses

In order to "read" skin signals and other cues from the environment, squid depend on their excellent eyesight. Squid eyes are huge in relation to their bodies. These big eyes gather enough light to let squid see at night and in the dark ocean depths.

Although eyesight is the squid's most important sense, these creatures can also smell, feel, and taste.

Squids use their nervous system to change the color of their skin. This helps the creatures blend in with their surroundings.

Smells are picked up by small pits beneath the eyes. Touching and tasting are done by the suckers. All of these senses help squid to get around in their underwater world.

The one sense squid do not use is hearing. Scientists believe squid probably cannot hear at all. However, they may be able to feel the movements of other animals in the water. This ability is much like hearing. It gives squid information about faraway creatures, just as sound gives humans information about barking dogs or other noisy objects that they cannot see. Between its eyesight and its other senses, a squid is well equipped for undersea life.

2

Making New Squid

It is not easy to track squid throughout their lives. For this reason, scientists do not know the average lifespans of most squid species. However, species that have been studied in detail seem to live between one and three years. It is reasonable to assume that less-studied species follow the same pattern.

A squid's short life includes a growing-up period and an adult period. As an adult, the squid's most important job is to **spawn**. This activity creates new squid to take the place of those that die. In this way, squid populations around the world stay steady and strong.

Early Life

A squid begins its life when it hatches from an egg. The newborn squid looks just like an adult, but it is much smaller. It may measure less than one-twentieth of an inch from end to end. The little squid becomes part of the plankton, a layer of tiny creatures floating near the ocean's surface. The squid eats any of these creatures it can catch. The longer the squid lives, the bigger and stronger it becomes.

Squid are not the only hungry creatures in the plankton. Tiny octopuses, sea stars, eels, and other creatures also live there, and they will eat the little squid if they can. Because of this, the plankton

These eggs belong to a type of squid found in the waters around Japan. A ruler has been placed underneath them to show how tiny the eggs are.

stage is a dangerous time. About half of all baby squid are eaten within their first week of life, and more will die within the next couple of weeks.

Some squid get lucky. They manage to survive for a few weeks. After this period, a squid is big enough to leave the plankton. However, it is still too small to protect itself from fish and other larger **predators**. Therefore it joins with thousands of other little squid to form enormous schools. The sheer size of these schools protects the young squid. Because there are so many targets for hungry predators to attack, individuals have a better chance of survival than they would have on their own.

Sometime between six months and one year of age, squid become adults. Some species stick together after they reach adulthood and form large schools throughout their lives. Other species go their separate ways. Solitary species will rejoin only when it is time to spawn.

Spawning

In most species, spawning takes place at set times of the year. Some species will spawn anywhere. Others have traditional spawning grounds. Squid may swim a long way to reach these areas.

When squid are ready to spawn, they start looking for mates. Most of the looking is done by the male, who must work hard to attract a female. He does this by swimming near the female and flashing

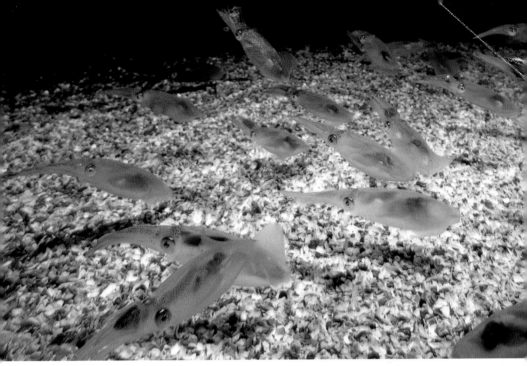

Several of the squid in this small school have fish in their arms. Squid have a better chance of survival in a school than they would on their own.

different skin patterns at her. Sometimes several males chase the same female, all flashing. The skin patterns tell the female that the males want to mate. In response, the female flashes back. Using her coloration, she tells the males whether she is interested and if so, which male she prefers.

After a female chooses her mate, the two squid wind their arms and tentacles together. They hold tightly to each other. The male then uses one of his arms to put a sperm packet into the female's mantle cavity. In some species, additional sperm packets are put into a special cavity below the female's mouth. The female will keep the packet until she is ready to lay her eggs.

Spawning Frenzies

Huge numbers of squid may gather for mass spawning sessions. These events are sometimes called "spawning frenzies." During a spawning frenzy, the ocean surface churns as millions of squid jet back and forth. This process may last several days. Males swarm around females, all competing for attention. Those that succeed grab the female from underneath and leave their sperm packets.

These squid are ready to mate. The male squid is below the female, and has attracted her attention by blinking his skin pattern.

Spawning frenzies are a frantic, confused time for squid. Small males sometimes take advantage of the confusion by "sneak mating" with females. This means that instead of competing for a female's attention, the male sneaks up on her and quickly leaves his sperm. A "sneaker" squid can zip in and out unnoticed while larger males are busy showing off for the female.

One way or another, most of the squid eventually mate. Eventually the males of some species leave the area. The males of other species weaken and start to die. At the same time, the females get ready to lay their eggs.

Laying Eggs

To lay eggs, the females of many species settle to the ocean floor. After reaching the bottom, the female starts passing egg clumps out of her funnel. The number of eggs in each clump varies from species to species. Some clumps contain just a few hundred eggs. Others may contain thousands of eggs. In all species, the eggs are fertilized by the male's stored sperm as they leave the body. The clumps begin hardening into rice-shaped packets as soon as they hit the seawater. The female uses a sticky string to attach each packet to the ocean floor.

Over a period of several days, the female lays many egg packets. She carefully attaches each new packet next to those she has already laid. She gets

A clump of squid eggs is attached to sea grass.
These eggs are probably less than two days old.

weaker and weaker as she does this job. Finally she dies. The eggs are now on their own.

Not all squid attach their eggs to the seafloor. Some species lay their eggs in the open water. They may attach their egg packets to seaweed or other floating objects, or they may simply release their eggs into the sea. The eggs are carried away as soon as they leave the female's body. They will drift with the ocean currents until they hatch.

A New Generation

Right after a squid's egg packets are laid, they are still soft. At this stage the packets may be eaten by fish, rays, and other predators. Very soon, however, the packets get so hard that they are difficult to chew open. They also seem to taste bad. Scientists have seen predators take mouthfuls of squid eggs, then spit them out. Sea stars will eat the eggs, but only if they cannot find any other prey.

Since squid eggs are seldom eaten, most of them complete their development. This process takes about four weeks. After this time, new squid emerge. The tiny creatures float away to start their life in the ocean. A new generation of squid has begun.

chapter 3

Eat or Be Eaten

Squid are **carnivores**, which means they eat other animals. Their favorite foods are fish, shrimp, crabs, and other squid. Always hungry, squid eat as many of these animals as they can catch.

Squid are perfectly suited to their hunting lifestyle. Between their sharp senses, speed, dangerous suckers, and other physical features, squid have many tools to help them feed and protect themselves.

Catching a Meal

To find prey, squid depend mostly on their eyesight. A squid looks around as it swims through the water. It tries to spot fish or other tasty creatures.

When it sees something it wants to eat, it jets forward. The squid moves quickly and silently so the prey will not notice its approach.

When the squid gets close enough, it strikes. It does this by thrusting its tentacles forward at lightning speed. As soon as the clubs touch the prey, suction power attaches them to its skin. In species with

The squid at the top is eating a fish, which it caught in a fast strike with its tentacles.

toothed suckers, the teeth also dig into the prey's flesh. This gives the squid an even better grip.

A squid's strike is incredibly fast. Some species can reach out and snatch prey in just thirty milliseconds (thousandths of a second). This process is much too quick to be seen by the human eye. It is also too quick for most prey animals to escape. When a squid strikes, it usually catches its meal.

Once an animal has been snagged by the tentacles, it is pulled back toward the squid's body. The squid then grabs the prey with its eight arms. Hundreds of suckers hold the prey tight as the squid prepares to eat.

Squid can become dinner for other hungry squid. Here a group of five squid feed on a smaller squid.

Eating

The squid uses its arms to move the prey toward its mouth. The mouth is hidden on the underside of the head between the arms and tentacles. The outer part of the mouth is a sharp, curved beak much like that of a parrot. The squid uses its strong beak to tear bite-sized chunks of flesh from its prey's body. At this time the squid may also inject a paralyzing poison into the prey so it cannot struggle.

After the squid has ripped its prey's flesh, it uses an organ called the **radula** to yank the food into its mouth. The radula is like a hard tongue lined with rows of backward-pointing hooks. These hooks spear the food and pull it into the squid's throat.

Once inside the throat, the food moves to the squid's stomach to be digested. The swallowing process can be dangerous. The throat of a squid passes right through its brain, so an extra big bite could damage the brain. A squid can even die if its food is too big. But this does not happen often.

Squid Defenses

Squid are not just powerful hunters. They are also one of the ocean's most important prey animals. Fish, whales, sharks, birds, and seals are just a few of the sea creatures that depend on squid for their survival.

Because squid are in constant danger of being eaten, they have developed many tricks to keep themselves safe. Hiding is the squid's main defense.

These animals use their incredible color-changing abilities to make themselves hard to spot. Some species also lurk near the ocean floor or in deep water during the daytime, when they can be easily seen by predators. When night comes, the squid rise from the bottom to hunt under cover of darkness.

If hiding fails and a squid is seen by a predator, it switches to a different strategy. It uses powerful water blasts to run away. A fleeing squid may flash different colors across its body to confuse its pursuer. Some species may also leap out of the water. Flying squid, as these species are called, can glide a hundred feet or more through the air before splashing back into the ocean.

Squid have another defense called "inking." Inside the squid's mantle is a sac that contains dark brown ink. This ink is called **sepia**. If a squid becomes frightened, it can blow a cloud of sepia from its funnel. The sepia cloud startles a predator. It also blocks the predator's view and may affect its sense of smell. While the predator is confused, the squid jets away to safety.

Glowing Squid

Some deepwater squid are **bioluminescent**. This means that parts of their bodies glow in the dark. A few bioluminescent squid use their glowing abilities to help them hunt. Some squid, for example, use their glow as a natural searchlight so they can see when it is dark. Other squid have glowing areas at

Bioluminescent squid use the glow from their bodies to help them hunt in the dark.

the tips of their arms. They use these bright spots to lure fish and other prey.

Most bioluminescent squid do not use their glow as a hunting tool. They use it for self-defense. Some squid, for example, flash bright lights at predators to startle and confuse them. Some can blast glowing ink into a predator's face. And some squid use bioluminescence to camouflage themselves. Lights on the bottom of the body, for example, can make a squid "disappear" when seen from below. The squid fades into the broken sunlight on the ocean's surface and becomes almost impossible to see.

Squid and Humans

A squid's defenses cannot protect it from one of its major enemies: humans. In some parts of the world, squid flesh (often called **calamari**) is an important part of people's daily diets. Commercial fishermen pull huge numbers of squid from the oceans each year to satisfy this market. Even more squid are used by fishermen as bait each year.

Still, squid and octopuses together account for only about 1 percent of the sea animals caught each year by humans. Whales, sharks, and other predators eat many more squid than people do. In fact, squid may be one of the few edible species in the oceans that are not being overfished by humans. As a result, squid populations around the world are healthy. For now, at least, there are plenty of squid in the sea.

chapter 21

The Giant Squid

In the depths of the ocean, a huge animal roams. The beast is over fifty feet long, with wriggling arms and enormous eyes. It has never been captured alive, but it definitely exists. Its remains sometimes wash up on beaches or become tangled in fishermen's nets. This mysterious creature is known simply as the giant squid.

What Is the Giant Squid?

Giant squid belong to the family **Architeuthis**. Scientists are not sure whether this family contains just one squid species or several. More than a dozen different types of Architeuthis have been named, but some scientists think there is only one

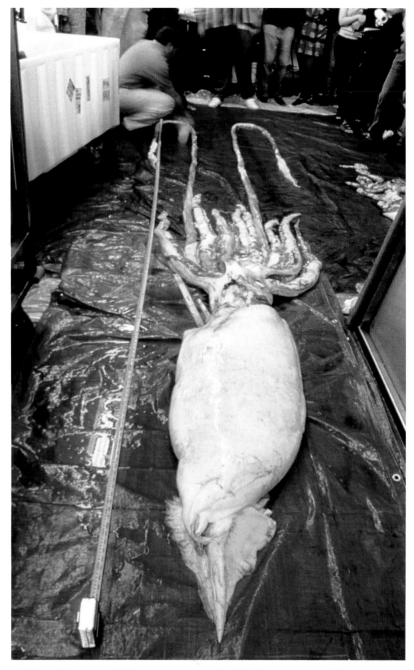

This giant squid was netted off the coast of New Zealand. It is twenty-five feet long and weighs two hundred and fifty pounds.

kind of giant squid that can be found all around the world. Most specimens have come from New Zealand, Newfoundland, and Norway, but giant-squid bodies have also been seen near Australia, the British Isles, the Bahamas, Florida, and many other places.

No one knows exactly how big giant squids may get. The largest body ever found measured fifty-seven feet from tip to tip. But many scientists believe that these animals may grow much larger—perhaps as long as one hundred feet. These scientists say it is very unlikely that humans accidentally found the biggest squid in the entire world. They are sure that larger ones exist.

Even a fifty-seven footer, however, is a huge and terrifying creature. A squid of this size would be the fifth-largest animal in the world. (Four types of whales grow longer.) It would have eyes the size of volleyballs, arms as thick as a human leg, and vast suckers ringed with knifelike teeth. If a person were caught by this monstrous squid, he or she would have no chance of survival.

Squid Sightings

Whalers have known about giant squid for centuries. Ancient whalers often saw their remains in whales' stomachs. Undigested tentacles and huge, sharp beaks proved that giant squid existed.

Every now and then an actual squid appeared instead of just its remains. One early sighting occurred in 1861 off the coast of the Canary Islands. On November 30 of that year, a ship's crew spotted a huge red squid floating at the ocean's surface. The squid's mantle and head alone were at least sixteen feet long. The crew tried to catch the squid by looping a rope around its body, but they accidentally pulled it to pieces as they hauled it toward the ship. They did, however, manage to grab part of the squid's mantle and bring it back to land. This finding gave scientists a chance to study pieces of Architeuthis.

Since the Canary Islands incident, hundreds of giant-squid bodies or parts have been found. In recent years most giant squid have been pulled up by

The remains of this giant squid washed up on a beach in Chile. It is more than forty feet long.

fishermen near Tasmania and New Zealand. All of these squid were netted at depths between one thousand and four thousand feet. This fact gives scientists an important clue about the habits of the giant squid.

Mortal Enemies

In their deep ocean homes, giant squid probably behave much like other squid do. It is likely that they spend most of their time swimming around, looking for food. With their huge bodies, strong arms, and powerful beaks, giant squid are dangerous predators indeed.

Like their smaller relatives, however, giant squid are not just predators. They may also become prey. The giant squid's main enemy is the sperm whale, which may reach sixty feet in length. Sperm whales dive thousands of feet to reach the giant squid's territory. Once there, they spend up to an hour searching for prey. They grab and swallow any giant squid they find.

A large sperm whale can easily eat a small giant squid. A really big squid, though, will fight to defend itself. The squid wraps its arms and tentacles around the attacking whale's head. It uses its toothed suckers to tear the whale's flesh, and it bites the whale with its sharp beak. If the squid is strong enough, it may drive its enemy away. The squid leaves round sucker scars on the whale's body as a reminder of the battle.

Sometimes a sperm whale pulls a giant squid to the ocean's surface. Doing this gives the whale an advantage, since it can breathe air and the squid cannot. The two huge creatures thrash the water into foam as they fight. More often than not, the squid loses the battle and becomes a meal.

Many people have reported seeing battles between giant squid and sperm whales. It is hard to say whether these reports are true. Most of them were written centuries ago when people could not take photographs to back up their stories. But even today, sperm whales' heads often bear large circular scars. These scars are evidence of the battles that rage between the whales and their mortal enemies the giant squid.

In Search of the Giant Squid

In recent years scientists have tried hard to find living giant squid. Two major expeditions were led by Dr. Clyde Roper in 1997 and 1999. Roper took teams to New Zealand's Kaikoura Canyon, which is the home of many sperm whales. Roper reasoned that where sperm whales lived, so did their favorite prey.

Roper's main goal on both of his expeditions was to film giant squid underwater. To do this, he sent down submarines and cameras on cables. He even used suction cups to attach video cameras to sperm whales' heads. He hoped that when the whales hunted for squid, the cameras would record all the action. Unfortunately Roper's methods did

As technology improves, perhaps scientists will learn more about the giant squid.

not succeed. The scientist never managed even to spot any giant squid, let alone film them. Roper's crew did catch some tiny Architeuthis babies, but they all died within a few hours.

Roper's experiences show just how hard it is to track giant squid. Underwater cameras and submarines can see only a tiny fraction of the ocean world. Even if there are millions of giant squid, the chances of finding and filming one are very small. For this reason, most investors do not want to give money to squid-hunting crews. This makes it hard for Roper and others to arrange new expeditions.

But scientists will surely try again. The giant squid is too interesting to ignore forever. And in the future, better technology might make it easier to find this mysterious animal. It seems certain that one day—perhaps soon—scientists will at last come face-to-face with the giant squid.

Glossary

Architeuthis: The scientific name of the giant squid.

bioluminescent: Able to glow in the dark.

calamari: Squid used as food for humans.

camouflage: Disguise. Squids camouflage themselves by changing their skin color to blend into the background.

carnivore: Any animal that eats only other animals.

cephalopod: An animal whose feet are attached directly to its head.

chitin: A hard substance similar to human fingernails.

chromatophores: Color cells in the skin that can get bigger or smaller, thereby changing a squid's color.

funnel: A flexible tube in the squid's head. Water, ink, and eggs are pushed out through the funnel.

giant axons: The squid's long nerves.

invertebrate: Any animal that does not have a backbone.

mantle: The arrowlike part of a squid's body.

mantle cavity: An open space just beneath the outer skin of the mantle.

pen: A hard, bonelike object to which the squid's muscles are attached.

predator: Any animal that hunts other animals.

radula: The squid's rough tongue.

sepia: Brown ink that can be squirted at predators.

siphon: See "funnel."

spawn: To release eggs or sperm.

suckers: Round suction disks that line the squid's arms and tentacle clubs.

tentacles: Thin, extralong arms ending in flat, sucker-bearing clubs.

For Further Exploration

Books

Brian Innes, *Water Monsters.* Austin, TX: Raintree Steck-Vaughn, 1999. This book describes sightings of mysterious underwater creatures, including the giant squid.

James Martin, *Tentacles: The Amazing World of Octopus, Squid, and Their Relatives.* New York: Crown Publishers, 1993. Learn a little bit about the entire cephalopod family in this easy-to-read book.

Websites

The Cephalopod Page (is.dal.ca/~ceph/TCP). This scientific site includes lots of cephalopod information, pictures, and links.

The Squid Page (www.mindspring.com/~erica/squid). Play squid games, send squid-mail to a friend, watch a dancing squid, and more.

Videos

Sea Monsters: Search for the Giant Squid. Questar, Inc., 2000. This National Geographic video follows Dr. Clyde Roper's expedition to Kaikoura Canyon, New Zealand.

Index

picture credits

Cover Photo: © James B. Wood
© Army/AFP/Getty Images, 36
© Roger T. Hanlon, 12, 21, 24, 27, 28
Chris Jouan, 14
© Reuters NewMedia Inc./CORBIS, 34
© Jeffrey L. Rotman/CORBIS, 39
© James B. Wood, 7, 10, 13, 16, 19, 22, 31

Kris Hirschmann has written more than ninety books for children. She is the president of The Wordshop, a business that provides a wide variety of writing and editorial services. She holds a bachelor's degree in psychology from Dartmouth College in Hanover, New Hampshire. Hirschmann lives just outside of Orlando, Florida, with her husband, Michael, and her daughter, Nikki.